WAR HAWGS: A-10s of the USAF

Also by Don Logan
Rockwell B-1B: SAC's Last Bomber
The 388th Tactical Fighter Wing At Korat Royal Thai Air Force Base 1972
Northrop's T-38 Talon
Northrop's YF-17 Cobra
Republic's A-10 Thunderbolt II
The Boeing C-135 Series: Stratotanker, Stratolifter, and other Variants
General Dynamics F-111 Aardvark
ACC Bomber Triad: the B-52, B-1B and B-2
USAF F-15 Eagles
F-4 Phantom IIs of the USAF Reserve and Air National Guard

WAR HAWGS

A-10s of the USAF

Don Logan

Schiffer Military History
Atglen, PA

Acknowledgments

I would like to thank Brian C. Rogers and Tom Kaminski for their help in the research for this book. Craig Brown supplied the Unit patches used in this book.

I would like to thank the Operations Personnel and the Public Affairs Offices of the A-10 Units worldwide without whose help I could not have put this book together.

I would also like to thank the following individuals who provided photographs included in this book: Brett Sweetman (25th Fighter Squadron and 355th Fighter Squadron photos), Darrell Walton (355th Fighter Squadron photos), Brian Ogawa (355th FS Mission Marking photos), Kevin Jackson (Eglin A-10 C), Stu Freer (81st Fighter Squadron Photos), Nigel Lulham (81st Fighter Squadron Photos), and David F. Brown (Bollen Range photos with Fall foliage background). Additional photographs were provided by Brian C. Rogers and Nate Leong. All photos without a credit line were taken by the Author.

The Author

After graduating from California State University-Northridge with a BA degree in History, Don Logan joined the USAF in August 1969. He flew as an F-4E Weapon Systems Officer (WSO), stationed at Korat RTAFB in Thailand, flying 133 combat missions over North Vietnam, South Vietnam, and Laos before being shot down over North Vietnam on July 5, 1972. He spent nine months as a POW in Hanoi, North Vietnam. As a result of missions flown in Southeast Asia, he received The Distinguished Flying Cross, The Air Medal with eleven oak leaf clusters, The POW Medal, and The Purple Heart. After his return to the U.S., he was assigned to Nellis AFB where he flew as a rightseater in the F-111A. He left the Air Force at the end of February 1977.

In March of 1977 Don went to work for North American Aircraft Division of Rockwell International, in Los Angeles, as a Flight Manual writer on the B-1A program. He was later made Editor of the Flight Manuals for B-1A #3 and B-1A #4. Following the cancellation of the B-1A production, he went to work for Northrop Aircraft as a fire control and ECM systems maintenance manual writer on the F-5 program. In October of 1978 he started his employment at Boeing in Wichita, Kansas as a Flight Manual/ Weapon Delivery manual writer on the B-52 program. He retired from Boeing as Lead Editor of the B-52 Flight and Weapon Delivery Manuals in February 2002. He has continued his writing on military aircraft subjects and aircraft photography.

Book design by Robert Biondi and Don Logan.

Copyright © 2007 by Don Logan.
Library of Congress Catalog Number: 2006929192.

Printed in China.
ISBN: 978-0-7643-2586-1

We are always looking for people to write books on new and related subjects. If you have an idea for a book, please contact us at the address below.

Published by Schiffer Publishing Ltd.
4880 Lower Valley Road
Atglen, PA 19310
Phone: (610) 593-1777
FAX: (610) 593-2002
E-mail: Info@schifferbooks.com.
Visit our web site at: www.schifferbooks.com
Please write for a free catalog.
This book may be purchased from the publisher.
Please include $3.95 postage.
Try your bookstore first.

In Europe, Schiffer books are distributed by:
Bushwood Books
6 Marksbury Ave.
Kew Gardens, Surrey TW9 4JF
England
Phone: 44 (0)20 8392-8585
FAX: 44 (0)20 8392-9876
E-mail: info@bushwoodbooks.co.uk
www.bushwoodbooks.co.uk
Free postage in the UK. Europe: air mail at cost.
Try your bookstore first.

Contents

THE A-10 .... **8**
The A-10C – Precision Engagement Upgrade Program 10

AIR COMBAT COMMAND (ACC) **14**
Davis Monthan AFB – 355th Wing 14
    354th Fighter Squadron – Bulldogs 19
    357th Fighter Squadron – Dragons 25
    358th Fighter Squadron – Lobos 31

Nellis AFB – U.S. Air Force Air Warfare Center 40
    422d Test and Evaluation Squadron 41
    66th Weapons Squadron 44

Pope AFB – 23rd Fighter Group 58
    74th Fighter Squadron – Flying Tigers 62
    75th Fighter Squadron – Tiger Sharks 67

U.S. AIR FORCES EUROPE (USAFE) **72**
Spangdahlem AB, Germany – 52nd Fighter Wing 72
    81st Fighter Squadron – Panthers

PACIFIC AIR FORCES (PACAF) **80**
Eielson AFB, Alaska – 354th Fighter Wing 80
    355th Fighter Squadron – Falcons

Osan AB, Korea – 51st Fighter Wing 96
    25th Fighter Squadron – Dragons

AIR FORCE RESERVE COMMAND (AFRC) **104**
Barksdale AFB – 917th Wing 104
    47th Fighter Squadron – Terrible Termites

NAS JRB New Orleans – 926th Fighter Wing 128
706th Fighter Squadron – Cajuns

Whiteman AFB – 442d Fighter Wing 144
303d Fighter Squadron – KC Hawgs

AIR NATIONAL GUARD (ANG) **160**
Connecticut ANG, Bradley Airport – 103rd Fighter Wing 160
118th Fighter Squadron – Flying Yankees

Idaho ANG, Boise Air Terminal – 124th Wing 174
190th Fighter Squadron

Maryland ANG, Martin State Airport, Baltimore – 175th Wing 180
104th Fighter Squadron

Massachusetts ANG, Barnes Airport – 104th Fighter Wing 194
131st Fighter Squadron

Michigan ANG, W.K.Kellogg Airport, – 110th Fighter Wing 206
172d Fighter Squadron

Pennsylvania ANG, NAS JRB Willow Grove – 111th Fighter Wing 212
103d Fighter Squadron – Fightin' 103d

APPENDICES
CURRENT A-10 OPERATING UNITS 222

CURRENT TAIL CODES 222

BRAC 2005 A-10 IMPACT 223

GLOSSARY 224

Introduction

This stand-alone pictorial book is intended to be both a complement and a supplement to my first A-10 book, *Republic's A-10 Thunderbolt II: A Pictorial History*. This book bridges the gap between 1997, when my first A-10 book was published, and summer 2006 with current A-10 photographs and information. In taking photographs I personally visited ten of the twelve continental U.S. bases operating A-10s. The only bases I was not able to visit were the Idaho ANG at Boise, Idaho and the Michigan ANG at Battle Creek, Michigan. I also photographed A-10s on the Nellis AFB range complex, Bollen Range at Fort Indiantown Gap near Harrisburg Pennsylvania, Razorback Range at Fort Chaffee near Fort Smith Arkansas, and Claiborne Range at Fort Polk near Alexandria, Louisiana. The three bases outside the forty-eight contiguous Untied States bases; Eielson AFB in Alaska, Spangdahlem AB in Germany, and Osan AB in South Korea are represented by photographs from other photographers.

The A-10

A-10 MARKINGS

Two units have applied faces to their A-10s, the 23rd Fighter Group "Flying Tigers" have their traditional tiger mouth and eye that dates back to the P-40s flown by the Flying Tigers of the American Volunteer Group in China at the beginning of World War II. The 47th Fighter Squadron, 917th Wing Air Force Reserve Command at Barksdale AFB, have a "hawg" face painted on all their A-10s. As a general rule, A-10s do not have nose art, instead, they have applied art to the inside of the boarding ladder door on the left side of the fuselage. When the ladder is stowed and the door is closed the art is no longer visible. Examples of ladder door art are included throughout this book.

Many units have aircraft marked as Commander's Aircraft, also called unit Flagships. These aircraft have special tail markings. There are Command aircraft for Wing (WG or FW), Operations Groups (OG), and Squadrons (FS or WPS). Normally the unit number and type is applied between the tail code and the serial number. Some units shade the tail code letters with white or light gray edging. Some units have aircraft whose serial numbers which contain part of the unit number. In that case, the serial number numbers that match the unit numbers are highlighted.

A-10 Production

The first YA-10 (serial number 71-1369) was assembled at the Fairchild Republic Farmingdale, Long Island, New York plant. It was shipped by C-5 to Edwards AFB. It made its initial test flight at Edwards AFB, California on May 10, 1972. The second prototype (71-1370) first flew on 21 July 1972. The flight-tests continued until mid-October.

Delivery of the six pre-production A-10As (listed officially on Air Force inventories as YA-10As) began in February 1975. In addition to the six DT&E aircraft, there were two airframes built for static and fatigue testing. The second airframe was tested in a separate test lab area in the Farmingdale, New York plant.

On July 31, 1974 DoD released $39 million to allow Fairchild to proceed with production of 52 A-10s, subject to the provision that contract options to procure a smaller quantity would be kept open. Three months later, the military qualification tests were competed on the TF-34-100 engines, and they were cleared for use in production A-10A aircraft. The production A-10A was quite similar to the six pre-production aircraft.

Production continued through aircraft buys in 1975, 1976, 1977, 1978, 1979, 1980, 1981, and 1982. A total of 715 A-10s were manufactured. Making up the 715 aircraft were:

2 prototypes,
6 pre-production, and
707 production aircraft

The 715 A-10s had the following serial numbers:

71-1369 - 71-1370 Prototypes (2)
73-1664 - 73-1673 Pre-production (6)
(1670 - 1673 cancelled)
75-0258 - 75-0279 Production (22)
75-0280 - 75-0309 Production (30)
76-0512 - 76-0554 Production (43)
77-0177 - 77-0276 Production (100)
78-0582 - 78-0725 Production (144)
79-0082 - 79-0243 Production (144)
80-0140 - 80-0283 Production (144)
81-0939 - 81-0998 Production (60)
82-0646 - 82-0665 Production (20)

Like many military aircraft, the official name of the A-10 - "The Thunderbolt II" is rarely used. Instead the nickname of Warthog, or simply HOG or HAWG, is more commonly used, hence the title of this book "WAR HAWGS".

Only one two-seat aircraft was made. 73-1664, the first of the pre-production aircraft, was modified to the two-seat configuration. This modification was not a pilot training aircraft but rather was modified to test the two-seat Night/All Weather Attack configuration. The two-seat aircraft was eventually designated as YA-10B.

The A-10's simple construction techniques kept tooling costs down. Standard conical or flat shapes were used for skin surfaces with few compound curves. This simple design kept the need for elaborate manufacturing tooling to a minimum. The aircraft had a large number of interchangeable components that could be used on the left and right side of the airframe. The vertical fins, rudders, main landing gear, wing panels, fuselage sections, and pylons were interchangeable. The A-10 was the first Air Force twin-engined aircraft designed with neutral engines; that is, the engine is assembled in a single configuration that can be installed in either the right or left nacelle. The advantages of this commonality are obvious; with fewer different parts the documentation and simplified spares support is reduced. A fully loaded A-10A can reach a maximum speed of 450 knots or 518 miles per hour. Typical loiter speed of the aircraft in a combat profile was around 390 to 400 miles an hour. Slower speeds were more desirable in attacking armor targets.

PRECISON ENGAGEMENT UPGRADE PROGRAM

The Precision Engagement (PE) Upgrade Program for the A-10 includes enhanced precision target engagement capabilities. This upgrade allows the deployment of precision weapons such as JDAM (Joint Direct Attack Munition) and Wind Corrected Munitions Dispenser (WCMD), as well as enabling an extension of the aircraft's service life to 2028. Improvements will include: two new multifunction cockpit displays, situational awareness data links, digital stores management system, 1760 weapons data bus, Litening AT or Sniper XR targeting pod for precision-guided weapons and helmet-mounted sighting system. Lockheed Martin Systems Integration is prime contractor for the Program. Low rate initial production of the upgrade started in 2004. Up to 125 A-10 aircraft are to be upgraded by 2007. Lockheed Martin was awarded a contract for the integration of the Sniper XR targeting pod on the A-10 as part of the PE Upgrade Program. This Modification Program was considered extensive enough to redesignate the modified aircraft from A-10A/OA-10A to A-10C/OA-10C.

COCKPIT

The modified cockpit is equipped with a new "heads-up" display, which is used for targeting and weapon aiming, a Have-Quick secure radio communications system, inertial navigation and a tactical air navigation (TACAN) system. The aircraft are also to be fitted with a Terrain Profile Matching System (TERPROM). The pilot is equipped with night-vision goggles and also the infrared imaging display of the AGM-65 Maverick.

A-10C COCKPIT

(Kevin Jackson)

WEAPONS

The aircraft has eleven stores pylons. There are three pylons under the fuselage and the loads can be configured to use either the centre-line pylon or the two flanking fuselage pylons. For weapon guidance, the aircraft can be fitted with Pave Penny laser guidance/electronic support measures, pod installed on the right side fuselage pylon. Each wing carries four stores pylons: three outboard and one inboard of the wheel fairing.

The A-10 can carry up to ten AGM-65 Maverick air-to-surface missiles. The Raytheon Maverick missile uses a variety of guidance systems, including imaging infrared guidance and warheads, including a high-penetration conical-shaped charge warhead. The A-10 can also carry the Sidewinder air-to-air missile, which is an all-aspect short-range missile with maximum speed over Mach 2.

The 30mm GAU-8 cannon, designed for destroying armor is mounted on the centerline of the fuselage and fires at rates of 2000-4000 rounds per minute. The gun is mounted on the aircraft centerline so that it will not impart a yaw to the aircraft when fired. The GAU-8 cannon's trajectory is almost flat, making it deadly at a range of 4000 feet, and capable of knocking out a tank at 6000 feet with a well-placed shot, or able to destroy lightly armored vehicles at two miles.

The Northrop Grumman Litening ER (Extended Range) targeting pod has been successfully integrated on the A-10A and will also be used on the A-10C. Litening ER features a 640 x 512 pixel thermal imager, CCD TV, laser spot tracker/rangefinder, IR marker and laser designator.

The new Lockheed Martin Sniper targeting pod is also being integrated on the A-10C. The wedge-nosed, long-range targeting pod features a mid-wave FLIR System, a laser spot-tracker, laser marker, CCD television camera and a combat-identification capability. The 440lb. Sniper pod has a 30cm diameter. The pod's wedge-shaped nose is made of highly durable sapphire which is transparent to visible and infrared wavelengths. The unique design gives Sniper a semi-low-observable characteristic.

(Kevin Jackson)

A-10C TEST AIRCRAFT

81-0989   (USAF)

81-0989   (USAF)

81-0989   (USAF)

81-0989 FIRST A-10C TEST AIRCRAFT (Kevin Jackson)

82-0658 SECOND A-10 TEST AIRCRAFT

78-0704 FIRST A-10C IN MARKINGS OF AN OPERATIONAL UNIT

Air Combat Command (ACC)

355th Wing

Davis Monthan AFB, Arizona

After several years of combat operations with F-105s in Southeast Asia, the 355th TFW reactivated at Davis-Monthan AFB, Arizona in July 1971. Initially equipped with A-7Ds, the Wing received its first A-10 in March 1976. The 355th became a Tactical Training Wing on October 1, 1979 after dropping its operational commitment in April 1979. The 355th's unit designations were modified to Fighter Wing and Fighter Squadrons on October 1, 1991. The 355th Wing has three A-10 Squadrons assigned; the 354th, 357th, and 358th. The 358th FS is the A-10

Pilot FTU (Formal Training Unit) and as such is responsible for training all new A-10 pilots. BRAC 2005 has not implemented any changes to the A-10s of the 355th Wing. Assigned A-10 strength remains at 66 aircraft.

A-10s of the 355th have worn nearly all of the camouflage schemes seen on A-10s and are presently painted in the standard two grays scheme. The black **DM** tail code represents the base name, **D**avis-**M**onthan.

81-0939 355th WING COMMANDER'S AIRCRAFT AT BAGRAM, AFGHANISTAN

(USAF)

355th WING ASSIGNED A-10s

78-0650	79-0196	80-0169	80-0236
78-0651	79-0198	80-0173	80-0246
78-0652	79-0201	80-0176	80-0270
78-0657	79-0202	80-0179	80-0278
78-0670	79-0209	80-0186	80-0280
78-0671	79-0210	80-0190	81-0939
78-0673	80-0141	80-0195	81-0941
78-0684	80-0142	80-0203	81-0942
78-0706	80-0146	80-0204	81-0943
78-0712	80-0147	80-0206	81-0948
79-0167	80-0150	80-0207	81-0950
79-0168	80-0151	80-0210	81-0961
79-0174	80-0155	80-0211	81-0974
79-0178	80-0159	80-0212	81-0982
79-0188	80-0162	80-0215	82-0648
79-0190	80-0167	80-0216	82-0662
79-0195	80-0168	80-0235	82-0663

The list of serial numbers at the front of each section is made up of serial numbers of aircraft that were assigned to the given unit in the time period between 1995 and 2006. They do not necessarily represent the aircraft assigned at the time of the writing of this book.

81-0939 355th WING COMMANDER'S AIRCRAFT

81-0939

79-0198

79-0198 12th AIR FORCE COMMANDER'S AIRCRAFT

80-0142 357th FS WEST COAST A-10 DEMONSTRATION AIRCRAFT

The West Coast A-10 Demonstration Team is currently assigned to Davis Monthan AFB flying aircraft 80-0142, 80-0168, 80-0173, and 81-0939 marked as the 355 Wing CC (Commander's) aircraft. In addition to the Wing Commander's Aircraft (81-0939) each of the Fighter Squadrons supplies an aircraft to the West Coast A-10 Demo Team. The Demo Team aircraft have a white tail cap with a yellow and blue sword, and the Air Force Wings symbol on the tail between the tail code and the ACC Emblem.

80-0168 354th FS WEST COAST A-10 DEMONSTRATION AIRCRAFT

80-0168 354th FS WEST COAST A-10 DEMONSTRATION AIRCRAFT

80-0173 358th FS WEST COAST A-10 DEMONSTRATION AIRCRAFT

80-0173 358th FS WEST COAST A-10 DEMONSTRATION AIRCRAFT

(Kevin Jackson)

354th Fighter Squadron
BULLDOGS

The 354th TFS flew A-7s between July 1971 and April 1, 1979. The 354th remained a "paper" squadron, inactivating on April 30, 1982 without having converted to A-10s. Like the 333rd FS, the squadron reactivated with OA-10s and served the 602nd ACW only from activation on November 1, 1991, until returning to the 355th Wing in May 1992. After returning to the 355th Wing, the 354th moved to McChord AFB, Washington, on January 5, 1993. The squadron tail band (retained from the 602nd) was dark blue, with BULLDOGS in white script and a cartoon bulldog on a yellow disc.

During October of 1994 the 354th returned to Davis-Monthan. Replacing the 333rd FS, the 354th inherited their red tail band adding a white bulldog head and "BULLDOGS" in black script. By 1997 the tail band had been changed back to blue edged in yellow, with a yellow bulldog head centered in the band. The 354th FS emblem is displayed in black outline on the left side of the fuselage aft of the cockpit with the 355th Wing emblem in black outline on the right side. The only exception to this is aircraft 80-0168, the 354th FS's West Coast Demo Team aircraft which has full color 354th FS and 355th Wing emblems.

81-0943 354th FIGHTER SQUADRON COMMANDER'S AIRCRAFT

78-0652

78-0657

78-0671

78-0671

78-0712

79-0167

79-0188

79-0201

79-0209

80-0181

80-0211

80-0236

80-0278

354th Fighter Squadron

The 357th TFS activated with A-7s in July 1971. It was redesignated the 357th TFTS on 1 July 1976, and began A-10 transition in April 1979. On October 1, 1991, the squadron became the 357th Fighter Squadron (FS). The original squadron marking was a yellow fin tip with black lightning bolt, but, by 1990, this design was modified to a yellow fin band edged in black with a black dragon's head centered in the band. The 357th FS emblem is displayed in full color on the left side of the fuselage aft of the cockpit with the 355th Wing emblem in full color on the right side.

78-0706 357th FIGHTER SQUADRON COMMANDER'S AIRCRAFT

78-0673

79-0196

79-0196

80-0141

80-0147

80-0280 & 81-0950

80-0147

80-0150

80-0212

80-0270

80-0280

81-0948

358th Fighter Squadron LOBOS

An A-10 student pilot is seen here on approach chased by his instructor on his left wing. A second student and instructor pair is behind them in the final turn. Because there are no two seat pilot training A-10s, flight training in the A-10 is accomplished with the student pilot in one A-10 chased by the instructor pilot in a second A-10.

The 358th TFS activated on June 1, 1972, became the 358th TFTS on January 1, 1976, and the 358th FS on October 1, 1991. A-10s arrived in January 1978. The 358th is the A-10 pilot training squadron for the USAF.

The first tail marking for the 358th was a bright green fin cap, later bordered by a white band. By late 1989, the revised marking was a green tail stripe. In early 1990 the fin band was again revised, this time to black with a wolf's head. The present tail band black edged in red with a wolf (lobo) head centered in the band. The 358th FS emblem is displayed in full color on the left side of the fuselage aft of the cockpit with the 355th Wing emblem in full color on the right side.

78-0651

78-0670

79-0174

79-0174

80-0151

80-0151

80-0151

80-0155

80-0159

80-0186

80-0187

80-0187

80-0206

80-0210 & 80-0155

80-0210

80-0216

80-0216

80-0235

80-0235

82-0662

82-0662

U.S. AIR FORCE WARFARE CENTER
NELLIS AFB, NEVADA

Established as the Air Warfare Center in October 1995, the US Air Force Warfare Center (USAFWC) was renamed in October 2005. The center, which is located at Nellis AFB, Nevada, manages advanced aircrew training and integrates many of the Air Force's test and evaluation requirements. It is an outgrowth of the USAF Tactical Fighter Weapons Center, established in 1966, which concentrated on the development of forces and weapons systems that were specifically designed for tactical air operations in conventional (non-nuclear) war.

The USAFWC uses the Nellis Air Force Range Complex, occupying about three million acres of land, the largest such range in the United States, and another five-million-acre military operating area which is shared with civilian aircraft. The Center also uses the range complex at Eglin AFB, Florida that adds even greater depth to the Center's capabilities, providing over water and additional electronic expertise to the Center. The Air Warfare Center oversees operations of the 57th Wing and 99th Air Base Wing at Nellis AFB and the 53rd Wing at Eglin AFB, Florida.

422nd Test and Evaluation Squadron

The 422nd Test & Evaluation Squadron (TES) was originally designated the 422nd Fighter Weapons Squadron (FWS). The squadron provides the continual refinement of combat techniques and tactics, including weapon release profiles, maneuvering, and electronic warfare tactics. It was assigned to the 57th Wing from October 15, 1969 until transferring to the 53rd Wing at Eglin AFB on October 1, 1996.

The 422nd Fighter Weapons Squadron began flying A-10s in the late 1970s. Tactical Air Command (TAC) directed the 422nd (1) to conduct operational tests and evaluations of fighter weapons systems and (2) to develop tactics for employment of those systems in combat. This mission led the 422nd to develop the JAWS camouflage schemes and evaluate the first Charcoal Lizard camouflages. The 422nd became a Test and Evaluation Squadron under the Deputy Commander for Tactics and Test on December 30, 1981.

Presently, the 422nd TES is a composite squadron and reports to the 53rd Wing via the 53rd Test and Evaluation Group (TEG) at Eglin AFB Florida, and operates OT tail coded aircraft from Nellis AFB. It conducts operational tests of A-10, F-22, F-15C, F-15E, F-16C and HH-60G hardware and software enhancements prior to release to the Combat Air Forces. The squadron develops and evaluates tactics to optimize the combat capability of these weapon systems in a simulated combat environment. The 422nd TES also develops and publishes new tactics for these aircraft. The results of these tests directly benefit aircrews in ACC, PACAF, and USAFE by providing them with operationally proven hardware and software systems.

79-0169	79-0171	79-0199
80-0242	82-0658	

79-0169

79-0169

79-0171

79-0199

82-0658 A-10C & 23 FG A-10 79-0157

82-0658 A-10C

66th Weapons Squadron

The 57th Fighter Weapons Wing (FWW) came to Nellis on October 15, 1969, replacing the USAF Tactical Fighter Weapons Center's 4525th FWW. (The 57th officially adopted the 4525th's bull's eye insignia in 1970.) There have been four subsequent redesignations: 57th Tactical Training Wing on April 1, 1977; 57th FWW (again) on 1 October 1980; 57th FW on October 1, 1991; and finally 57th Wing on June 15, 1993.

USAF Weapons School/ 66th Weapons Squadron

The first 57th A-10s arrived in October 1977, and were assigned to the 66th Fighter Weapons Squadron. They were used to train a cadre of weapons instructors from all A-10 units. The 66th FWS inactivated at the end of 1981, in a major reorganization of the USAF Fighter

Weapons School. The 66th's mission moved to the A-10 Division, USAF Fighter Weapons School. In June 1993, the school was renamed the USAF Weapons School (USAFWS) with the Weapons Instructor Course still belonging to the A-10 Division. On February 3, 2003 the 66th was activated as the 66th Weapons Squadron replacing the A-10 Division, USAFWS.

A-10 tail markings are made up of the **WA** (**W**eapons & **A**rmament) tail code with a black and yellow checkerboard band. BRAC 2005 has not implemented any changes to the A-10s of the 66th Weapons Squadron. Assigned A-10 strength remains at 10 aircraft.

80-0185	80-0200	80-0204
80-0229	80-0234	81-0946
81-0958	81-0977	
75-0301 (Maintenance Trainer)		

81-0977 WEAPONS SCHOOL COMMANDER'S AIRCRAFT

80-0229 66th WEAPONS SQUADRON COMMANDER'S AIRCRAFT

80-0229 66th WEAPONS SQUADRON COMMANDER'S AIRCRAFT

80-0185

80-0185

80-0200

80-0200

80-0200

81-0946

81-0946

81-0958

81-0958

81-0958

A-10s carry infrared flares used to protect the aircraft from heat seeking missiles and chaff used to protest the aircraft from radar directed missiles. The flare and chaff dispensers are located beneath the wing tips and behind the main landing gear. The aircraft in the above photo has just released an infrared flare.

81-0958

81-0958

81-0958

81-0958

81-0958

81-0958

81-0977

81-0977

81-0977

81-0977

81-0977

81-0977

81-0977

81-0977

81-0977

23rd Fighter Group
FLYING TIGERS

Pope AFB, North Carolina

Pope's A-10 unit, the 23rd Fighter Group, carries the lineage of the Flying Tigers of World War II. Though it is presently based at Pope AFB, North Carolina, it is assigned to the 4th Fighter Wing at Seymour-Johnson AFB, North Carolina as a "geographically separate unit". The 23rd FG is a tenant organization to the 43rd Airlift Wing (C-130Es), the primary flying unit at Pope AFB.

The 23rd FG has two flying squadrons, the 74th and 75th Fighter Squadrons operating under the 23rd Operation Support Squadron. The A-10s at Pope carry the **FT** tail code (derived from **F**lying **T**igers).

As a result of BRAC 2005, the 36 A-10s of the 23rd Fighter Group will be moving to a new Wing to be set up at Moody AFB, Georgia.

23rd FIGHTER GROUP ASSIGNED A-10s

78-0596	79-0157	80-0144	81-0947
78-0597	79-0159	80-0172	81-0953
78-0598	79-0162	80-0175	81-0964
78-0600	79-0179	80-0180	81-0967
78-0674	79-0186	80-0194	81-0990
78-0679	79-0189	80-0208	82-0657
78-0688	79-0192	80-0223	82-0660
78-0697	79-0204	80-0228	82-0661
79-0135	79-0206	80-0229	82-0664
79-0138	79-0213	80-0252	
79-0139	79-0223	80-0277	
79-0141	80-0140	80-0282	

80-0194 EAST COAST DEMO TEAM - 23rd FIGHTER GROUP COMMANDER'S AIRCRAFT

The East Coast A-10 Demonstration Team is currently assigned to Pope AFB flying aircraft 79-0223, 80-0194, 80-0208, and 80-0223. The Demo Team aircraft carry the Air Force Wings symbol on the tail between the tail code and the ACC Emblem.

80-0208 EAST COAST DEMO TEAM - 23rd FIGHTER GROUP COMMANDER'S AIRCRAFT

79-0223 EAST COAST DEMO TEAM - 23rd FIGHTER GROUP COMMANDER'S AIRCRAFT

80-0223 EAST COAST DEMO TEAM - 23rd FIGHTER GROUP COMMANDER'S AIRCRAFT

80-0223 EAST COAST DEMO TEAM - 23rd FIGHTER GROUP COMMANDER'S AIRCRAFT

79-179 EAST COAST DEMO TEAM AIRCRAFT

79-0179 EAST COAST DEMO TEAM AIRCRAFT

74th Fighter Squadron (Flying Tigers)

The 74th Tactical Fighter Squadron (TFS) was the first squadron in the 23rd TFW to begin transition to A-10s. Transition began in September 1980 when the Squadron and Wing were located at England AFB, Louisiana flying A-7Ds. The 74th TFS was redesignated 74th Fighter Squadron (FS) on November 1, 1991 and inactivated at England AFB, Louisiana on February 15, 1992. It was subsequently reactivated as part of the 23rd FG at Pope AFB on June 15, 1993.

In the early 1980s, the 74th tail tops were blue with white stars. These tail tops were later changed to blue with a white lightning bolt. The 74th FS emblem is displayed in black outline on the left side of the fuselage aft of the cockpit with the 23rd FG emblem in black outline on the right side.

81-0964 DESERT STORM Mi-8 HELICOPTER KILLER

Now assigned to the 74th FS, 81-0964 was assigned to the 511th TFS while deployed to Saudi Arabia during Desert Storm. On 15 February 1991 964 being flown by Capt. Todd Sheehy was credited with shooting down an Iraqi Mi-8 helicopter. 964 wears a green star below the cockpit as credit for the kill.

79-0135

79-0192

79-0206

79-0206

80-0140

80-0228

80-0228

80-0228

80-0228 still had its Operation Iraqi Freedom Mission markings on its ladder door when photographed here at Pope AFB in October 2004.

81-0947

81-0967

81-0964 DESERT STORM Mi-8 HELICOPTER KILLER

81-0990

75th Fighter Squadron (Tiger Sharks)

The squadron tail tops were black and white checkerboards. The 75th FS emblem is displayed in black on the left side of the fuselage aft of the cockpit with the 23rd FG emblem in black on the right side.

78-0600

78-0600

78-0600

78-0688

79-0129

79-0157

79-0213

80-0180

80-0180

82-0660

82-0660

82-0664

82-0664

U.S. Air Forces Europe
(USAFE)

PANTHERS
81ST FIGHTER SQUADRON

**81st Fighter Squadron
52nd Fighter Wing**

Spangdahlem AB, Germany

(Stu Freer/Fighter Control)

The 510th FS brought its A-10s to Spangdahlem, Germany, from RAF Bentwaters, England, in January 1993. Just over a year later, on 25 February 1994, the 510th FS moved (minus personnel and equipment) to Aviano, Italy to fly F-16s. Spangdahlem's A-10 unit then became the 81st FS. For most of the first year, **SP** (**SP**angdahlem) tail code was the only unit-applied markings. Black fin caps with white lightning bolts were added and a black panther head painted on both engine nacelles was seen at different times. The 81st FS recently repainted the fin caps yellow with black lightning bolts. The 52nd Fighter Wing emblem is painted in black on both sides of the fuselage aft of the cockpit.

52nd FIGHTER WING A-10s

81-0951	81-0963	81-0983	81-0992
81-0952	81-0966	81-0984	82-0649
81-0954	81-0976	81-0985	82-0650
81-0956	81-0978	81-0988	82-0654
81-0962	81-0980	81-0991	82-0656

The 81st FS Panthers deployed to Bagram AB, Afghanistan in support of Operation Enduring Freedom between June and November 2003 flying over 4500 hours and 1,400 sorties in support of Coalition Forces.

The Panthers operated in a completely blacked-out environment with the living area, maintenance, taxi, take-off, and landing safely all under extremely hazardous NVG conditions.

80-0281 81st FIGHTER SQUADRON COMMANDER'S AIRCRAFT (Nigel Lulham)

81-0956 (Nigel Lulham) 80-0281 (Nigel Lulham)

Standard 81st Fighter Squadron Tail cap is yellow with a centered black lightning bolt. The Squadron Commander's aircraft has a black panther head instead of the lightning bolt.

80-0281 81st FIGHTER SQUADRON COMMANDER'S AIRCRAFT

(Nigel Lulham)

81-0951   **(Stu Freer/Fighter Control)**

81-0951   **(Nigel Lulham)**

81-0951   **(Stu Freer/Fighter Control)**

81-0952
(Stu Freer/Fighter Control)

81-0956
(Nigel Lulham)

81-0976
(Stu Freer/Fighter Control)

81-0980

(Stu Freer/Fighter Control)

81-0983

(Nigel Lulham)

81-0985

(Nigel Lulham)

81-0988 (Nigel Lulham)

81-0991 (Stu Freer/Fighter Control)

81-0992 (Stu Freer/Fighter Control)

81-0991 & 81-0992

(Stu Freer/Fighter Control)

82-0650

(Stu Freer/Fighter Control)

82-0654 **(Stu Freer/Fighter Control)**

82-0656 **(Stu Freer/Fighter Control)**

Pacific Air Forces
(PACAF)

354th Fighter Wing
Eielson AFB, Alaska

(Darrell Walton)

On October 1, 1981 Alaskan Air Command (AAC) activated the 343rd Composite Wing at Eielson AFB, Alaska. To prepare for A-10s, the 343rd took control of the 25th TASS with 0-2s (at Eielson) and the 18th TFS with F-4Es (at Elmendorf). The 18th received two A-10s in December and moved to Eielson on January 1, 1982. The 343rd was redesignated as a TFW on June 8, 1984. Despite plans to convert the 25th TASS to OA-10s, the 25th inactivated on September 1, 1989. On August 9, 1990, the Alaskan Air Command was redesignated the Eleventh Air Force and assigned to Pacific Air Forces)PACAF. On July 1, 1991, the 343rd was redesignated the 343rd Wing and the 18th, which had begun converting to F-16s in March, became the 18th

FS. At that time the 11th TASS (activated on May 10, 1991) took control of seven OA-10s. On August 20, 1993 the 343rd Wing was inactivated and was replaced by the 354th FW. The 354th Fighter Wing had been the A-10 Wing at Myrtle Beach AFB, South Carolina until that based closed in March 1993.

As a result of the Base Realignments and Closure (BRAC) Commission's 2005 decisions the 354th Fighter Wing will close and its eighteen A-10 aircraft will be reassigned as follows: twelve aircraft to the new wing being formed at Moody AFB, Georgia, three aircraft to the 917th Wing at Barksdale and three aircraft to backup inventory.

355th Fighter Squadron   (USAF)

With the activation of the 354 Fighter Wing, the 355th FS (one of the A-10 Squadrons from Myrtle Beach AFB) replaced the 11th TASS on August 20, 1993. Initially the replacement of wing and squadron insignia was the only change in unit markings. The first gray A-10 was painted on base in February 1994. The 355th FS initially used the red tail stripe of 11th TASS. The stripe color was changed to black during 1997. The PACAF emblem is on both engine nacelles on some aircraft. The PACAF emblem was originally on the vertical fin above the tail code, but was moved when the stars of the Big Dipper were added to the outside of the vertical fin.

354th FIGHTER WING A-10s

78-0701	80-0178	80-0240	81-0970
79-0172	80-0189	80-0254	81-0979
79-0187	80-0197	80-0259	81-0995
79-0207	80-0220	80-0272	81-0997
80-0149	80-0226	81-0944	
80-0172	80-0238	81-0969	

81-0995 354th FIGHTER WING COMMANDER'S AIRCRAFT   (Brian C. Rogers)

81-0979 354th OPERATIONS GROUP COMMANDER'S AIRCRAFT   (Darrell Walton)

81-0979 354th OPERATIONS GROUP COMMANDER'S AIRCRAFT

(Darrell Walton)

(Darrell Walton)

81-0979 354th OPERATIONS GROUP COMMANDER'S AIRCRAFT (Darrell Walton)

81-0979 354th OPERATIONS GROUP COMMANDER'S AIRCRAFT (Darrell Walton)

80-0238 355th FIGHTER SQUADRON COMMANDER'S AIRCRAFT, BAGRAM, AFGHANISTAN FALL 2004 (USAF)

80-0238 355th FIGHTER SQUADRON COMMANDER'S AIRCRAFT

During the 355th FS Deployment to Bagram mission tallies were added on the left side under the cockpit. The markings consisted of bullet silhouettes for strafe kills, laser guided bomb silhouettes for bombing missions and Maverick silhouettes for missile launches.

80-0238 355th FIGHTER SQUADRON COMMANDER'S AIRCRAFT

(Brian Ogawa)

(Darrell Walton)

80-0238 355th FIGHTER SQUADRON COMMANDER'S AIRCRAFT (USAF)

80-0238 355th FIGHTER SQUADRON COMMANDER'S AIRCRAFT (Brett Sweetman)

80-0238 355th FIGHTER SQUADRON COMMANDER'S AIRCRAFT (Brett Sweetman)

Because of the extreme cold during the winters at Eielson, heated "taxi through" hangars are used for A-10 flight operations. 80-0238 is seen here shutting down in its hangar after flight.

MISSION MARKINGS FROM FALL 2004 DEPLOYMENT TO BAGRAM, AFGHANISTAN

80-0172

(Brian Ogawa)

(Darrell Walton)

80-0187

(Brian Ogawa)

(Darrell Walton)

80-0197

(Brian Ogawa)

(Darrell Walton)

80-0220

(Brian Ogawa)

(Darrell Walton)

80-0226

(Brian Ogawa)

(Darrell Walton)

80-0240

(Brian Ogawa)

(Darrell Walton)

80-0272

(Brian Ogawa)

(Darrell Walton)

81-0944

(Brian Ogawa)

PROCEED WITH CAUTION

(Darrell Walton)

81-0970

(Brian Ogawa)

(Darrell Walton)

79-0172 (Brett Sweetman)

79-0187 (Darrell Walton)

79-0207

80-0149 (Darrell Walton)

80-0189 (Darrell Walton)

80-0197 (Darrell Walton)

80-0197 (Darrell Walton)

80-0220 (Darrell Walton)

80-0220 (USAF)

80-0226 (USAF)

80-0259 (Darrell Walton)

81-0944 (Darrell Walton)

81-0969 (Darrell Walton)

81-0969 (Darrell Walton)

81-0970

(Brett Sweetman)

EIELSON AFB RAMP

(USAF)

81-0997

(Brett Sweetman)

(Darrell Walton)

25th Fighter Squadron 51st Fighter Wing

Osan AB, Korea

The 25th TFS relocated to the newly rebuilt Suwon AB, Korea from Kadena AB, Okinawa without personnel or equipment on February 1, 1981. The personnel arrived in March 1982, and the A-10s soon followed. The 25th TFS's parent unit was the 51st Composite Wing (Tactical) at Osan. The 51st became a TFW on July 1, 1982. When Seventh Air Force activated on September 8, 1986, it assumed control of the 51st and all other Korean-based PACAF assets from 5th Air Force. The 25th, planning F-16 conversion, transferred the A-10s to the 19th TASS in August 1989 and the 19th relocated to Osan in October 1990. The 25th continued to exist on paper, even borrowing a few F-16s as part of the conversion plans, until its inactivation on July 31, 1990.

All 51st Wing A-10s were delivered in European I camouflage. The original tail code was OS (for Osan, the wing headquarters). By early 1984 the codes became SU (for Suwon), at the end of summer of 1990 the tail code had changed back to OS

25th Fighter Squadron (Assam Dragons)

The 25th was reactivated on October 1, 1993 and the 19th TASS was inactivated as a result of the Air Force's attempts to retain as many of its historically prominent units as downsizing would permit. At the same time the 51st TFW was redesignated the 51st Fighter Wing. The Assam Dragons markings consist of **OS** tail codes (for **OS**an), a galloping horse on both sides of the fuselage and a yellow and green checked tail cap.

51st FIGHTER WING A-10s

78-0685	80-0165	80-0241	80-0283
79-0183	80-0177	80-0243	81-0959
79-0185	80-0192	80-0244	81-0973
79-0211	80-0213	80-0245	82-0651
80-0153	80-0217	80-0247	82-0652
80-0163	80-0224	80-0251	
80-0164	80-0239	80-0253	

(Brett Sweetman)

79-0183 51st OG COMMANDERS AIRCRAFT

(Brett Sweetman)

All the images in the 25th Fighter Squadron Chapter were taken by Brett Sweetman at Osan AB, Korea.

79-0183 51st OG COMMANDERS AIRCRAFT

(Brett Sweetman)

80-0251 25th FS COMMANDERS AIRCRAFT

(Brett Sweetman)

80-0251 25th FS COMMANDERS AIRCRAFT

(Brett Sweetman)

80-0253 25th FS COMMANDERS AIRCRAFT

(Brett Sweetman)

78-0685 & 80-0283

(Brett Sweetman)

78-0685

(Brett Sweetman)

79-0211   **(Brett Sweetman)**

80-0153   **(Brett Sweetman)**

80-0153   **(Brett Sweetman)**

80-0165 (Brett Sweetman)

80-0213 (Brett Sweetman)

80-0224 (Brett Sweetman)

80-0239   **(Brett Sweetman)**

80-0241   **(Brett Sweetman)**

81-0971   **(Brett Sweetman)**

81-0971 (Brett Sweetman)

82-0651 (Brett Sweetman)

82-0652 (Brett Sweetman)

Air Force Reserve Command
(AFRC)

47th Fighter Squadron

Barksdale AFB, Louisiana

Barksdale's first A-10s arrived in June 1980 and were assigned to the 47th TFS, 917th Tactical Fighter Group (then a part of the 434th TFW at Grissom). The A-10s were the first front line aircraft ever delivered direct from the factory to AFRES. The 917th achieved wing status in 1987 when the 926th TFG at New Orleans was assigned. All units dropped the word Tactical from their designations on February 1, 1992. In October 1993 the 93rd Bomb Squadron activated with B-52Hs, giving the redesignated 917th Wing a combat capability never

before seen in the Reserves. On October 1, 1996, the 917th Wing again began training A-10 pilots. The 47th FS augments the 355th Wing at Davis-Monthan training A-10 pilots.

A-10s along with the B-52Hs of the 917th carry **BD** (**B**arks**D**ale) tail codes. The 47th A-10s fin cap marking is derived from their Squadron Patch. It is made up of the word "Dogpatchers" and a "Terrible Turnip Termite". Both go back to their heritage of Al Capp's Lil' Abner Cartoon characters. Each A-10 also has a nose art of an

Al Capp cartoon character applied to the left side of the fuselage just forward of the wing. The nose art location was dictated by the most notable marking on Barksdale A-10s which is the fearsome warthog head painted in black and white on the nose of each of its A-10s.

917th WING A-10s

79-0094	79-0145	79-0150	79-0155
79-0095	79-0146	79-0151	80-0171
79-0105	79-0147	79-0152	
79-0120	79-0148	79-0153	
79-0142	79-0149	79-0154	

As a result of Base Realignments and Closure (BRAC) commission's 2005 decisions the 917th Wing will be adding nine more aircraft, six from the 926th Fighter Wing closing at NAS JRB New Orleans, Louisiana, and three from the 354th Fighter Wing, closing at Eielson AFB, Alaska. This will bring the number of A-10 aircraft assigned to the 917th Wing to twenty four.

DOGPATCH CARTOON ART

79-0094

79-0095

79-0105

79-0120

79-0142

79-0145

79-0146

79-0147

79-0148

79-0149

79-0150

79-0151

PAPPY YOKUM

79-0152

MARRYIN' SAM

79-0153

EVIL EYE FLEEGLE

79-0154

BALD IGGLE

79-0155

SHMOOS

80-0171

This line-up of 47th FS A-10s was photographed at Hawgsmoke 2004 at Alexandria, Louisiana, the old England AFB, on 2 May 2004.

79-0153 MARRYIN' SAM 917th WING COMMANDER'S AIRCRAFT

79-0153 MARRYIN' SAM 917th WING COMMANDER'S AIRCRAFT

79-0153 MARRYIN' SAM 917th WING COMMANDER'S AIRCRAFT

79-0147 MAMMY YOKUM 47th FS COMMANDER'S AIRCRAFT

79-0147 MAMMY YOKUM 47th FS COMMANDER'S AIRCRAFT

79-0147 MAMMY YOKUM 47th FS COMMANDER'S AIRCRAFT

79-0147 MAMMY YOKUM 47th FS COMMANDER'S AIRCRAFT

79-0147 MAMMY YOKUM 47th FS COMMANDER'S AIRCRAFT

79-0094 OLD MAN MOSE **79-0094 OLD MAN MOSE**

79-0094 OLD MAN MOSE

79-0095 STUPEFYIN' JONES

79-0095 STUPEFYIN' JONES

Smoke from the GAU-8 30mm nose gun engulfs 79-0095 on a strafe pass at Claiborne Range.

79-0095 STUPEFYIN' JONES

79-0095 STUPEFYIN' JONES

79-0095 STUPEFYIN' JONES

79-0105 DAISY MAE

79-0105 DAISY MAE & 79-0148 MOONBEAM McSWINE

79-0105 DAISY MAE

79-0142 LONESOME POLECAT

79-0142 LONESOME POLECAT

79-0142 LONESOME POLECAT

79-0142 LONESOME POLECAT

79-0142 LONESOME POLECAT

79-0145 TINY YOKUM

79-0145 TINY YOKUM

79-0145 TINY YOKUM

79-0146 Li'L ABNER

79-0148 MOONBEAM McSWINE

79-0148 MOONBEAM McSWINE

79-0148 MOONBEAM McSWINE

This A-10 is seen releasing a BDU-33 Practice Bomb. The bomb weighing about 25 pounds simulates larger/heavier gravity bombs. It contains no explosive, only a spotting charge that makes a bright flash and puff of white smoke when it impacts the ground.

79-0148 MOONBEAM McSWINE

79-0149 GENERAL BULLMOOSE

79-0149 GENERAL BULLMOOSE

79-0150 WOLF GAL

79-0151 CAVE GAL

79-0151 CAVE GAL

79-0151 CAVE GAL

79-0151 CAVE GAL

79-0151 CAVE GAL

79-0151 CAVE GAL

79-0152 PAPPY YOKUM

79-0152 PAPPY YOKUM

79-0152 PAPPY YOKUM

79-0152 PAPPY YOKUM

79-0154 EVIL EYE FLEEGLE

79-0154 EVIL EYE FLEEGLE

79-0155 BALD IGGLE

79-0155 BALD IGGLE

80-0171 SHMOOS

706th Fighter Squadron CAJUNS

NAS JRB New Orleans, Louisiana

The 706th Fighter Squadron, 926th Fighter Wing, located at what is now referred to as Naval Air Station Joint Reserve Base New Orleans, Louisiana received its first A-10s in December 1981 and completed conversion in June 1982. The group was assigned to the 434th TFW at Grissom AFB, Indiana. It was transferred to the 442nd TFW at Richards-Gebaur AFB, Missouri on February 1, 1984, and to the 917th TFW at Barksdale on July 1, 1987. The 926th was activated for Desert Shield/Desert Storm on December 29, 1990. It was the only AFRES fighter unit called and the first Reserve fighter unit called up since the Korean War. The squadron deployed eighteen aircraft, including one loaned from the 47th TFS at Barksdale. The group was relieved from active duty on June 15. On February 1, 1992, the units dropped the word Tactical from their designations. The group transitioned to F-16s in October 1992.

On May 22, 1996, the 706th FS began their conversion back to the A-10 with the receipt of 78-655 and 80-237 from the 20th Fighter Wing at Shaw AFB.

The 47th FS at Barksdale provided 79-106 and 79-136 on June 17, 1996. The 926th Fighter Wing finished its transition back to the A-10 in early 1997.

The 706th Fighter Squadron teamed up with the 303rd Fighter Squadron from Whiteman AFB, Missouri, to deploy to Bagram Air Base, Afghanistan, as the 706th Expeditionary Fighter Squadron (EFS) in support of Operation Enduring Freedom. The deployment ran from April 4, 2002 through July 20, 2002 with the 706th flying over 500 sorties and more than 1200 hours in support of the U.S. and Coalition ground forces in Afghanistan.

The Base Realignments and Closure (BRAC) commission's 2005 decisions resulted in the closure of the 706th Fighter Squadron and 926th Fighter Wing. On March 15, 2006 the Unit's 15 aircraft were divided between the other AFRC A-10 Units with the 917th Wing at Barksdale AFB, Louisiana gaining six aircraft and the 442nd Fighter Wing at Whiteman AFB, Missouri gaining nine aircraft.

926th FIGHTER WING A-10s

78-0582	79-0111	79-0180	80-0237
78-0655	79-0121	79-0197	82-0653
78-0716	79-0134	80-0160	
79-0093	79-0136	80-0188	
79-0106	79-0144	80-0232	

The 926th Fighter Wing A-10 markings are similar to those applied to their F-16. The A-10 tail carries the **NO** (**N**ew **O**rleans) tail code. In keeping with the 926th use of Mardi Gras colors, the new red tail stripe has yellow fleurs-de-lis painted in the stripe. A smaller stripe below the main stripe indicates the maintenance flight to which the aircraft is assigned with purple for A flight and green for B flight. "CAJUNS' in script is on both sides of the fuselage aft of the cockpit.

79-0582

79-0582

79-0582

79-0655

79-0655 & 82-653

79-0106

79-0106

79-0106

79-0106

79-0111

79-0111

79-0111

79-0111

79-0111

79-0121

79-0121

79-0136

79-0136

79-0136

It should be noted in the above photo the serial number on the tail has been applied in error with the AF and year group following the last three instead of in front where they belong.

In Spring 2005 79-0136 still had the temporary ladder door art that was added while deployed in Afghanistan.

79-0136

79-0144

79-0144

79-0197

79-0197

79-0197 & 79-0111

79-0197 & 79-0111

79-0197

79-0197

79-0197 & 79-0111

80-0160

80-0160

80-0160

80-0188

80-0237

80-0237

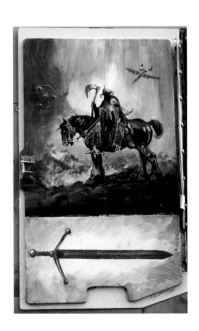

82-0653

79-0197, 79-0111, 78-0655, & 82-0653

79-0197, 79-0111, 78-0655, & 82-0653

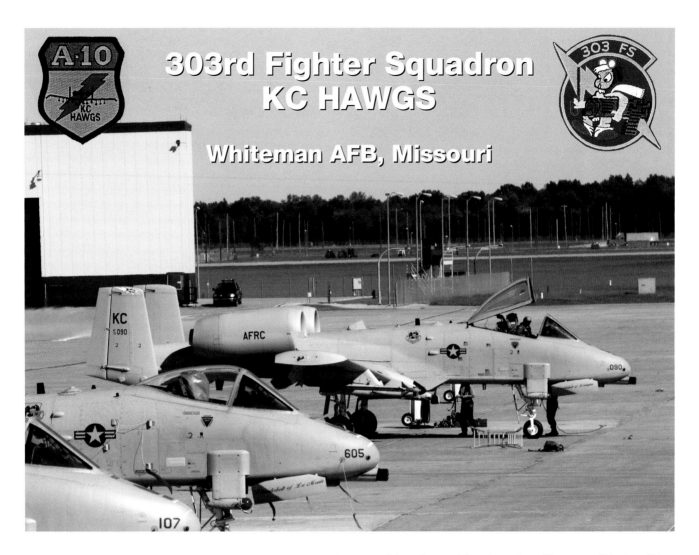

303rd Fighter Squadron
KC HAWGS

Whiteman AFB, Missouri

In October 1982 the 442nd Tactical Airlift Wing (TAW) became the 442nd Tactical Fighter Group (TFG) (with the 303rd TFS) based at Richards-Gebaur AFB, Missouri and began converting to A-10s; redesignation to the 442nd TFW occurred in February 1984. Tactical was dropped from unit designations on February 1, 1992. On June 11 and 12, 1994, the Wing moved to Whiteman AFB, Missouri sharing the ramp with the B-2s of the 509th Bomb Wing.

The 303rd Fighter Squadron teamed up with the 706th Fighter Squadron from Naval Air Station Joint Reserve Base, New Orleans, La., to deploy to Bagram Air Base, Afghanistan as the 706th Expeditionary Fighter Squadron (EFS) in support of Operation Enduring Freedom. The deployment ran from April 4, 2002 through July 20, 2002 with the 706th flying over 500 sorties and more than 1200 hours in support of the U.S. and Coalition ground forces in Afghanistan.

In February 2003, more than 300 members of the 442nd Fighter Wing were mobilized and directed to prepare for deployment in support of what would become Operation Iraqi Freedom. Initially, the 442nd was to be part of the planned "northern front" but as political realities in adjacent countries began to dictate resistance to that part of the plan, the Wing's expected deployment was put on hold. Finally the deployment order came at the very end of March and the airmen, with 12 A-10s, headed across the Atlantic. Arriving in Kuwait just as the battle in Iraq had progressed up from the south toward Baghdad, the wing's pilots, in their A-10s, participated in the Battle of Baghdad on their first day's sorties. A few days later the reservists received the order to forward deploy to an Iraqi airfield that had been all but destroyed in the first Persian Gulf War. They became the first U.S. Air Force fighter unit to forward deploy into Iraq and opened Tallil Air Base near the ancient Babylonian city of Ur as an Operation Iraqi Freedom coalition base of operations.

Not quite a month after their arrival at Tallil, the 442nd received the order to forward deploy to another Iraqi airfield; this time to Kirkuk AB. The A-10s landed at Kirkuk on April 29, and immediately began 24-hour flying operations. The Wing generated 1,164 combat sorties for a total of 3,360 flying hours with no combat casualties and no aircraft damages or losses.

A-10s of the 442nd arrived in standard European I camouflage, with the first repaints to the current Grays occurring in 1993. The tail code is **KC** (**K**ansas **C**ity); tail top markings consist of a yellow arrowhead overlaid with a white lightning bolt and a red "KC" with a front profile of an A-10 in gray on top. Each of the 442's A-10 is named after a town or city in the local area. The exceptions to this are 79-0122 "Thunderbolt of the 442d Fighter Wing", the Wing flagship; and 79-0123 "Thunderbolt of the 303d Fighter Squadron", the Squadron Flagship.

442nd FIGHTER WING A-10s

78-0605	THUNDERBOLT OF La MONTE
78-0631	THUNDERBOLT OF HIGGINSVILLE
79-0090	THUNDERBOLT OF LEXINGTON
79-0091	THUNDERBOLT OF KNOB NOSTER
79-0092	THUNDERBOLT OF CLINTON
79-0107	THUNDERBOLT OF CONCORDIA
79-0109	THUNDERBOLT OF SEDALIA
79-0110	THUNDERBOLT OF HOLDEN
79-0113	THUNDERBOLT OF BRANSON (Changed in 2005 from KANSAS CITY)
79-0114	THUNDERBOLT OF CLOUMBIA
79-0117	THUNDERBOLT OF CALIFORNIA
79-0118	THUNDERBOLT OF WARRENSBURG
79-0119	THUNDERBOLT OF WINDSOR
79-0122	THUNDERBOLT OF THE 422D FIGHTER WING
79-0123	THUNDERBOLT OF THE 303D FIGHTER SQUADRON
79-0164	THUNDERBOLT OF JEFFERSON CITY
80-0201	THUNDERBOLT OF COLE CAMP

As a result of the 2005 Base Realignment and Closure (BRAC) commission decisions the 442nd will be gaining an additional nine aircraft from the 926th FW at New Orleans.

442nd FW LADDER DOOR ART

79-0109

79-0113

79-0123

79-0122 THUNDERBOLT OF THE 442D FIGHTER WING

79-0122 THUNDERBOLT OF THE 442D FIGHTER WING

79-0122 THUNDERBOLT OF THE 442D FIGHTER WING

79-0122 THUNDERBOLT OF THE 442D FIGHTER WING

79-0123 THUNDERBOLT OF THE 303D FIGHTER SQUADRON

79-0123 THUNDERBOLT OF THE 303D FIGHTER SQUADRON

78-0605 THUNDERBOLT OF La MONTE

78-0631 THUNDERBOLT OF HIGGINSVILLE

78-0631 THUNDERBOLT OF HIGGINSVILLE

78-0631 THUNDERBOLT OF HIGGINSVILLE

78-0631 THUNDERBOLT OF HIGGINSVILLE

79-0090 THUNDERBOLT OF LEXINGTON

79-0091 THUNDERBOLT OF KNOB NOSTER

79-0091 THUNDERBOLT OF KNOB NOSTER

79-0092 THUNDERBOLT OF CLINTON

79-0107 THUNDERBOLT OF CONCORDIA

79-0107 THUNDERBOLT OF CONCORDIA

79-0107 THUNDERBOLT OF CONCORDIA

The B-2s of the 509th Bomb Wing are also located at Whiteman AFB, Missouri. The B-2s are always kept in hangars. Each B-2 has its own hangar like the one in the background in this photo.

79-0107 THUNDERBOLT OF CONCORDIA

79-0109 THUNDERBOLT OF SEDALIA

79-0109 THUNDERBOLT OF SEDALIA

79-0109 THUNDERBOLT OF SEDALIA

79-0109 THUNDERBOLT OF SEDALIA

79-0110 THUNDERBOLT OF HOLDEN

79-0110 THUNDERBOLT OF HOLDEN

79-0113 THUNDERBOLT OF BRANSON (Formerly KANSAS CITY)

79-0113 THUNDERBOLT OF BRANSON (Formerly KANSAS CITY)

79-0114 THUNDERBOLT OF COLUMBIA

79-0119 THUNDERBOLT OF WINDSOR

79-0164 THUNDERBOLT OF JEFFERSON CITY

Air National Guard
(ANG)

118th Fighter Squadron
103rd Fighter Wing

Connecticut ANG

In the summer of 1979, the 103rd TFG, 118th TFS (Flying Yankees) began converting from F-100s to A-10s at Bradley International Airport, Windsor Locks, Connecticut. In doing so, the 103rd became the first ANG unit to fly A-10s, and the first ANG unit to receive new tactical aircraft directly from the manufacturer. By 1991, the 103rd Tactical Fighter Wing was preparing to convert to F-16s, but the planned conversion was first delayed then canceled. Tactical was dropped from both designation in 1992 and the group was redesignated the 103rd Fighter Wing on October 1, 1995. A-10s remained in service with the 103rd into 2006. Connecticut Air Guard units dropped the term Tactical from their designations on March 16, 1992.

103rd FIGHTER WING A-10s

78-0586	78-0633	78-0707	81-0965
78-0613	78-0638	79-0084	82-0646
78-0615	78-0639	79-0108	
78-0621	78-0643	79-0165	
78-0625	78-0646	81-0960	

As a result of 2005 Base Realignment and Closure (BRAC) commission decisions the 103rd Fighter Wing and 118th Fighter Squadron will be inactivated and the A-10s will be transferred to the 188th Fighter Wing at Fort Smith, Arkansas replacing F-16s.

78-0621 80th ANNIVERSARY MARKINGS   **(103 FW)**

All 103rd A-10s were delivered in European I camouflage, but have now been repainted to the standard grays scheme. The Connecticut tail codes are **CT** (**C**onnecticu**T**). The units new tail markings appeared in 1995 and consisted of a black lightning bolt underlined with a blue stripe and topped by a yellow stripe. A 118 FS squadron patch has been applied to the left side of the aircraft. This patch is in full color on some aircraft.

For the 80th anniversary of the Flying Yankees, 78-0621 received special markings inspired by those carried on the Unit's P-51s during World War II. After the anniversary, the markings were modified slightly and retained on 78-0621 as the 103rd Fighter Wing Commanders Aircraft.

78-0621 80TH ANNIVERSARY MARKINGS   **(Nate Leong)**

78-0621 103rd FIGHTER WING COMMANDERS AIRCRAFT

78-0621 103rd FIGHTER WING COMMANDERS AIRCRAFT

78-0621 103rd FIGHTER WING COMMANDERS AIRCRAFT

78-0621 103rd FIGHTER WING COMMANDERS AIRCRAFT

LEFT ENGINE **RIGHT ENGINE**

The same emblems are applied to the fuselage sides of the other 103rd FW A-10s
The 118th FS Flying Yankee is on the left side and the Hawg's head 103rd Fighter
Wing emblem is on the right side.

78-0621 103rd FIGHTER WING COMMANDERS AIRCRAFT

78-0621 103rd FIGHTER WING COMMANDERS AIRCRAFT

78-0586

78-0586

These photos were taken at Hawgsmoke 2004, the A-10 gunnery meet occurring every other year. In 2004 the meet was rained out and as a result picked up the name Hawgwash 2004 instead of Hawgsmoke.

78-0586 carries the old 103rd Fighter Wing tail markings. The tail markings have been updated copying those used on 78-0621, the 80th Anniversary aircraft.

78-0586

78-0633

78-0633

78-0633

78-0633

78-0613

78-0613

78-0613

The three photos of 78-0613 on this page show the older markings, 78-0625 on the opposite page has the new 103rd FW markings.

78-0625

78-0625

78-0625

78-0638

78-0638   **78-0638**

Many of the 103rd FW's A-10s have retained ladder door art and mission markings from their combat deployments.

78-0643

78-0643

78-0643

78-0643

78-0646

78-0646   **78-0646**

78-0646

The 103rd FW has both 78-0646 and 82-0646. A small "8" has been added to the three digit number on both sides of the nose of 78-0626. A small "2" has been added in a similar manner on 82-0646

79-0165

81-0965

81-0965

81-0965

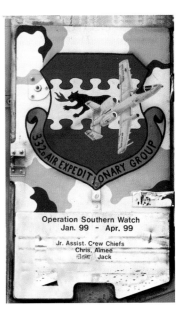

81-0965

81-0965

82-0646

82-0646

82-0646

82-0646

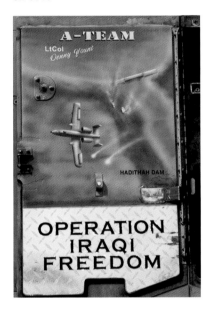

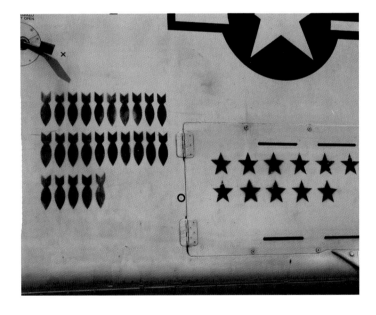

82-0646

190th FIGHTER SQUADRON, 124th WING IDAHO ANG

The 190th Fighter Squadron, 124th Wing, Idaho Air National Guard, located at Boise Airport/Gowen Field, outside Boise, Idaho began converting to the A-10 when their first aircraft arrived from the Massachusetts Air National Guard on March 20, 1996. The 190th sent their last F-4, (the last combat F-4 in the U.S. military), an F-4G Wild Weasel to AMARC at Davis-Monthan on April 20, 1996. Local area training in the A-10 began in August of 1996, with an operational commitment to begin in October 1997. In addition to the A-10 Squadron, the 189th Airlift Squadron is assigned to the 124th Wing and operates C-130s.

124th WING A-10s

78-0584	78-0624	78-0653	80-0191
78-0611	78-0627	78-0691	80-0218
78-0618	78-0629	78-0703	80-0250
78-0619	78-0634	79-0194	81-0955

As a result of the 2005 Base Realignments and Closure (BRAC) commission decision's the 124th Wing will pick up three more A-10s as the 103rd Fighter Squadron at NAS JRB Willow Grove, Pennsylvania closes. That will bring their total strength to 18 A-10s.

78-0618

(Brian C. Rogers)   (Brian C. Rogers)

A skull and crossed GAU-8 guns from the 190th Squadron patch is on both engine nacelles. The tail markings consist of an <u>ID</u> (<u>ID</u>aho) tail code and a tail with two mountain peaks and IDAHO in blue in red, blue, or white.

78-0618

78-0618

78-0618

78-0619

78-0619

78-0619

78-0624

78-0624

80-0218

80-0218

80-0250

80-0250

80-0250

104th Fighter Squadron Maryland ANG

Martin State Airport Baltimore, Maryland

Converting from A-37B Dragonflies, the 104th Tactical Fighter Squadron, 175th Tactical Fighter Group accepted its first A-10 at Warfield Air National Guard Base (ANGB), Martin State Airport, near Baltimore in October 1979. The unit received the first Low-Altitude Safety and Targeting Enhancement (LASTE) modified A-10s in the Guard and then used them to win the GUNSMOKE 91 air-to-ground bombing and gunnery competition. The term Tactical was dropped from unit designations in March 1992. The 175th Fighter Group was redesignated 175th Wing on October 1, 1995. Despite years of rumors about F-16 conversions, the 175th will continue to fly A-10s, adding three more A-10s coming from Willow Grove, Pennsylvania as a result of BRAC 2005.

The 104th Fighter Squadron, 175th Wing, accepted its first A-10 in October 1979. The unit received the first Low-Altitude Safety and Targeting Enhancement (LASTE) modified A-10s in the Guard and then used them to win the GUNSMOKE 91 air-to-ground bombing and gunnery competition. The term Tactical was dropped from unit designations in March 1992. Despite years of rumors about F-16 conversions, the 175th will continue to fly A-10s, adding three more A-10s coming from Willow Grove, Pennsylvania as a result of BRAC 20025.

In addition to the 104th FS, the 135th Airlift Squadron flying C-130Js is also assigned to the 175th Wing.

175th WING A-10s

78-0637	78-0682	78-0683
78-0693	78-0694	78-0702
78-0704	78-0705	78-0717
78-0718	78-0719	78-0720
79-0082	79-0086	79-0087
79-0088	79-0175	

The 104th FS A-10s tail code is **MD** (**M**arylan**D**). Their tail stripe is a modified version of the Maryland State Flag. The 175th Wing emblem is on the left engine nacelle and 104th FS emblem is on the right nacelle. The Pilot and Crew Chief names are contained in a stripe on the side of the aircraft under the windshield. The Baltimore Ravens insignia is applied in front of the stripe containing the names (see page 184).

79-0175 175 WING COMMANDER'S FLAGSHIP

79-0175 175 WING COMMANDER'S FLAGSHIP

78-0683

78-0683

78-0693

78-0693

78-0693

(David F. Brown)

The photos in this book with the autumn foliage background were taken by David F. Brown at Bollen Range, Ft. Indiantown Gap near Harrisburg, Pennsylvania 3-4 November 2005.

78-0694

78-0694

78-0694

78-0702

78-0702

78-0705

78-0717

78-0717

78-0717

78-0718

78-0719

78-0719

78-0719 here just before touchdown has the speed brakes deployed. Deploying speed brakes allows the pilot to fly with higher engine power settings, then is he decides he must abort the landing and go around he doesn't have to wait for the turbo fan engines to come up to speed all he has to do is decrease the drag on the aircraft by closing the speed brakes.

78-0719

78-0720

78-0720 hit a coyote on the runway while deployed to **DM AFB**. As a result "**COYOTE KILLER**" Nose Art was added to 78-0720.

79-0082

79-0086

(David F. Brown)

79-0087

79-0087

79-0087

79-0087

79-0087

79-0087

79-0087

79-0087 has just fired an infrared flare in the humid September morning air of Bollen Range.

79-0088

79-0088

The Massachusetts Air Guard converted from F-100D/ Fs to A-10s in July 1979. The 131st Fighter Squadron and 104th Fighter Wing are based at Barnes Municipal Airport, near Westfield, Massachusetts. These units were redesignated the 131st FS and 104th FG in March 1992 and the 104th Fighter Wing on October 1, 1995. Long rumored to be on the F-16 conversion list, the 104th remained in A-10s until the 2005 Base Realignment and Closure (BRAC) Commission recommended they give up their A-10s and receive F-15s from the 102nd Fighter Wing Massachusetts ANG based at Otis AFB on Cape Cod. The 2005 BRAC recommendations reassigned the A-10s of the 104th to the 188th Fighter Wing, Arkansas ANG at Fort Smith.

In 1995 the 104th deployed for six weeks to Operation Deny Flight and Deliberate Force, flying combat missions for the first time in the unit's history. In 1999 members of the 104th Fighter Wing were activated per a Presidential Selective Recall to deploy in support of NATO operations in Kosovo. One hundred fifty unit personnel deployed to Trapani, Italy to join members from A-10 units from Battle Creek, Michigan and Boise, Idaho.

The three units combined to form a joint expeditionary squadron that flew more than 400 combat sorties. In 2000 the unit deployed to Al Jaber Air Base, Kuwait in support of Air Expeditionary Force 7.Beginning in January 2003, the Wing deployed approximately 500 personnel in support of Operation Iraqi Freedom to an undisclosed location in Southwest Asia. There, the Wing became the lead unit of the 387th Air Expeditionary Wing, a combined unit comprised of 103rd Fighter Wing, Connecticut Air National Guard personnel and personnel from other units. The Wing sent 11 A-10s to the deployed location where they flew each day of the war, compiling a 98 percent Mission Effectiveness Rate with no combat losses or damage. Recognizing its service as the lead unit of the 387th Air Expeditionary Group, 410th Air Expeditionary Wing, in Operation Iraqi Freedom, the Wing was presented its seventh overall Air Force Outstanding Unit Award for the period from January 1 - May 21, 2003.

All 104th A-10s arrived in European I camouflage. Black **MA** (**MA**ssachusetts) tail code was applied. The ANG shield was stenciled in black above the right fuselage strake until 1981. A red tail band bearing five white stars and a black stripe was added by mid-1981. That year, a simple map of Massachusetts was added to each nacelle in camouflage colors. A 104 shows through the eastern part of the map in the background camouflage color. The map disappeared by the late 1980s. In 1989, a full-color Outstanding Unit Citation (with two oak leaves) was added to each side of the fuselage above the strakes. All MA A-10s are now in the standard two gray paint scheme and carry the 104th's red stripe with five white stars in a + pattern, three stars in the stripe with one star above and one star below the center star in the stripe. They carry a subdued 104th Fighter Wing emblem on the left engine nacelle and a subdued 131st Fighter Squadron emblem on the right engine nacelle.

Each of the 104th's A-10 carries the map of Massachusetts on the right side of the nose with the location and name of a city or town in the local area. Three of the Wing's aircraft have special markings. 79-0104 is the 104th Fighter Wing Commander's flagship and in addition to the name "City of Westfield" (the location of Westfield Barnes Airport) 79-0104 is also called "Spirit of Massachusetts" which is written in script on both engine nacelles. During 2004 the Boston Red Sox won the Baseball World Series and the New England Patriots won the National Football League Super Bowl. Since both these teams were from the State of Massachusetts, the 104th painted special Red Sox markings on 78-0644 and New England Patriots markings on 78-0626 to commemorate the events.

SERIAL NUMBER	CITY/TOWN NAME
78-0583	City of Chicopee
78-0612	City of Northampton
78-0614	City of Springfield (Springfield Falcons Hockey Team)
78-0616	Town of South Hadley
78-0626	Spirit of New England (Patriots Jet) was Town of Southwick
78-0630	City of Easthampton
78-0632	Town of Montgomery
78-0640	Town of West Springfield
78-0642	Town of Russell 131 Fighter Squadron Flagship
78-0644	City of Boston(Red Sox Jet) was Town of Huntington
78-0647	Town of Ludlow
78-0649	Town of Southampton
78-0659	Town of East Longmeadow
78-0696	Town of Agawam
79-0104	City of Westfield 104 Fighter Wing Flagship
79-0216	City of Holyoke
80-0166	Town of Palmer

As with many of the A-10 units, the 104th Fighter Wing has added art to the inside of the pilot's ladder door. Many of these images were painted by a local artist, Michael Dooney.

MICHAEL DOONEY 104TH FIGHTER WING LADDER DOOR ART

78-0614

78-0630

78-0649

78-0696

79-0104

79-0216

79-0104 CITY OF WESTFIELD 104 FIGHTER WING COMMANDER'S AIRCRAFT **(USAF)**

79-0104 CITY OF WESTFIELD 104 FIGHTER WING COMMANDER'S AIRCRAFT

79-0104 CITY OF WESTFIELD 104 FIGHTER WING COMMANDER'S AIRCRAFT

78-0642 TOWN OF RUSSELL 131 FIGHTER SQUADRON COMMANDER'S AIRCRAFT

78-0642 TOWN OF RUSSELL 131 FIGHTER SQUADRON COMMANDER'S AIRCRAFT

78-0642 TOWN OF RUSSELL 131 FIGHTER SQUADRON COMMANDER'S AIRCRAFT

78-0644 CITY OF BOSTON – RED SOX 2004 WORLD SERIES CHAMPIONS AIRCRAFT

78-0644 received special markings to celebrate the victory of the Boston Red Sox in the 2004 World Series.

78-0644 CITY OF BOSTON – RED SOX 2004 WORLD SERIES CHAMPIONS AIRCRAFT

78-0626 SPIRIT OF NEW ENGLAND – PATRIOTS SUPER BOWL CHAMPIONS AIRCRAFT

78-0626 received special markings com-memorating the victory of the New England Patriots in Super Bowl 38 (2004) and Super Bowl 39 (2005).

78-0626 SPIRIT OF NEW ENGLAND – PATRIOTS SUPER BOWL CHAMPIONS AIRCRAFT

78-0614 CITY OF SPRINGFIELD – SPRINGFIELD FALCONS HOCKEY TEAM AIRCRAFT

78-0614 CITY OF SPRINGFIELD – SPRINGFIELD FALCONS HOCKEY TEAM AIRCRAFT

78-0614 CITY OF SPRINGFIELD – SPRINGFIELD FALCONS HOCKEY TEAM AIRCRAFT

78-0583 CITY OF CHICOPEE

78-0612 CITY OF NORTHAMPTON   **(USAF)**

78-0632 TOWN OF MONTGOMERY

78-0616 TOWN OF SOUTH HADLEY

78-0649 TOWN OF SOUTHAMPTON

78-0632 TOWN OF MONTGOMERY

78-0640 TOWN OF WEST SPRINGFIELD

78-0647 TOWN OF LUDLOW

78-0649 TOWN OF SOUTHAMPTON

79-0216 CITY OF HOLYOKE

80-0166 TOWN OF PALMER

172nd FIGHTER SQUADRON, 110th FIGHTER WING, MICHIGAN ANG

The 172nd Tactical Air Support Squadron, 110th Tactical Air Support Group, officially converted from OA-37s to A-10s on October 1, 1991, receiving most of the aircraft from the 343 TFW in Alaska. The 110th is based at W.K. Kellogg Airport in Battle Creek, Michigan.

The units were redesignated the 172nd Fighter Squadron and 110th Fighter Group on October 16, 1991 in conjunction with the conversion to OA-10s. The 110th FG was redesignated the 110th Fighter Wing on October 1, 1995.

All aircraft were delivered from Alaska in European I camouflage. The first Ghost Grays aircraft was repainted in 1992. The present markings are **BC** tail codes (**B**attle **C**reek), and "Michigan" in black on the outside top of both vertical stabilizers, and "Battle Creek" and the silhouette of the state of Michigan on the engine nacelle.

As a result of the 2005 Base Realignments and Closure (BRAC) commission decisions the 110th Fighter Wing at Battle Creek is set to lose its flying mission. The BRAC commission directed that the A-10s of the 110th FW are to move to Selfridge ANG Base, replacing the F-16s of the 127th Wing.

110th FIGHTER WING A-10s

79-0129	80-0257	80-0265	81-0996
80-0221	80-0258	80-0267	81-0998
80-0222	80-0262	80-0269	
80-0255	80-0263	81-0975	
80-0256	80-0264	81-0994	

80-0221

80-0221

80-0221

80-0221

80-0222

80-0222

80-0222

80-0222

80-0256

80-0256

80-0258

During Operation Iraqi Freedom, 80-0258 was hit in the right engine by an Iraqi surface to air missile. It returned to base and after being repaired was returned to operational service.

80-0258

80-0258

80-0258

80-0262

80-0265

81-0994

103rd Fighter Squadron 111th Fighter Wing

The 103rd Tactical Air Support Squadron of the 111th Tactical Air Support Group, Pennsylvania ANG, operating from NAS Willow Grove, Pennsylvania, began converting to OA-10s in 1988. The unit's last OA-37 left in March 1989. The 111th was slated to go to Desert Storm, but the rapid end of combat obviated the need for an additional A-10 Forward Air Control unit. The flying units were redesignated the 103rd Fighter Squadron and the 111th Fighter Group on March 16, 1992. The 111th FG was redesignated as the 111th Fighter Wing on October 1, 1995.

The decisions made by the 2005 Base Realignments and Closure (BRAC) commission have directed the inactivation of the 111th FW and the 103rd FS dividing their A-10s between the Idaho, Maryland and Michigan Air National Guard Units.

The unit uses **PA** (**P**ennsylvani**A**) tail code. Though originally and for many years called "The Black Hogs", political correctness has forced the unit to change their nickname back to the "Fightin' 103rd". The present tail markings consist of "Philadelphia" in black script across the top with a blue stripe with a yellow-orange lightning bolt. The winged horse from the 103rd Squadron insignia is applied over the lightning bolt. The 103rd squadron insignia is applied to the left side of the aircraft with the 111th Fighter Wing insignia on the right side.

111th FIGHTER WING A-10s

78-0641	79-0219	80-0230	82-0647
78-0658	80-0152	80-0273	82-0659
78-0692	80-0184	80-0275	
79-0170	80-0196	81-0949	
79-0193	80-0214	81-0981	

82-0647

(David F. Brown)

80-0214 111th FIGHTER WING COMMANDER'S AIRCRAFT

80-0214 111th FIGHTER WING COMMANDER'S AIRCRAFT

80-0214 111th FIGHTER WING COMMANDER'S AIRCRAFT

78-0641

78-0658

78-0692

79-0170

(David F. Brown)

79-0219 & 80-0214

79-0219 & 80-0214

79-0219

79-0219

80-0152

80-0184

80-0196

80-0230

81-0949

81-0981

81-0981

81-0981

80-0273

The yellow and red rectangle on the left side of the ladder door's says **PENNSYLVANIA TERROR-IST HUNTING PERMIT** and the one on the right says **USA TERRORIST HUNTING PERMIT.**

82-0647

(David F. Brown)

82-0652

82-0659

Appendices

CURRENT A-10 OPERATING UNITS

Current A-10 Strength 2005:

	Total Active Inventory (TAI)* A-10s/OA-10	Primary Aircraft Inventory (PAI)* A-10/OA-10
Active Duty	129/75	108/54
ANG	76/26	72/18
AFRC	44/7	39/6
Total	249/108	219/78
Total both types	357	297

*TAI: Aircraft assigned to operating forces for mission, training, test, or maintenance. Includes primary, backup, and attrition aircraft.
*PAI: Aircraft assigned to meet primary aircraft authorization (PAA).

Active Duty USAF CONUS (Continental U.S.) Units

23rd Fighter Group, Pope AFB, North Carolina (component of the 4th Fighter Wing
Seymour-Johnson AFB, North Carolina)(to be moved to Moody AFB as a result of BRAC 2005)
57th Wing Nellis AFB, Nevada
355th Wing, Davis Monthan AFB, Arizona

U.S. Air Forces Europe (USAFE) Units

52nd Fighter Wing, Spangdahlem AB, Germany

Pacific Air Force (PACAF) Units

51st Fighter Wing Osan AB, Korea
354th Fighter Wing, Eielson AFB, Alaska (To be closed as a result of BRAC 2005)

Air Force Reserve Command Units

442nd Fighter Wing, AFRC, Whiteman AFB, Missouri
917th Wing, AFRC, Barksdale AFB, Louisiana
926th Fighter Wing, AFRC, NAS JRB New Orleans, Louisiana (To be closed as a result of BRAC 2005)

Air National Guard Units

103rd Fighter Wing, Bradley International Airport, Connecticut (To be closed as a result of BRAC 2005)
104th Fighter Wing, Westfield Barnes Airport, Massachusetts (A-10s replaced by F-15s as a result of BRAC 2005)
110th Fighter Wing, W.K. Kellogg Airport, Battle Creek, Michigan (Moved aircraft to Selfridge ANGB Michigan and replaced the F-16s of the 127th Fighter Wing as a result of BRAC 2005)
111th Fighter Wing, NAS JRB Willow Grove, Pennsylvania (To be closed as a result of BRAC 2005)
175th Fighter Wing, Martin State Airport, Baltimore, Maryland ANG
124th Wing, Boise Airport/Gowen Field, Idaho ANG
188th Fighter Wing , Fort Smith Regional Airport, Arkansas ANG (Added as a result of BRAC 2005)

Test and Support Units

46th Test Wing, Eglin AFB, Florida
53rd Wing, Eglin AFB, Florida
442nd Test and Evaluation Squadron, 53rd Test and Evaluation Group, Nellis AFB, Nevada
82nd Training Wing (TRW), Sheppard AFB, TX

CURRENT A-10 TAIL CODES

AK
355th Fighter Squadron, 354th Fighter Wing, Eielson AFB, Alaska

BC
172nd Fighter Squadron, 110th FW, W.K. Kellogg Airport, Michigan ANG

BD
47th Fighter Squadron, 917th Wing, AFRES, Barksdale AFB, Louisiana

CT
118th Fighter Squadron, 103rd Fighter Wing, Bradley International Airport, Connecticut ANG

DM
355th Wing Davis Monthan AFB, Arizona

ET
46th Test Wing, Eglin AFB, Florida

FT
23rd Fighter Group, Pope AFB, North Carolina

ID
190th Fighter Squadron, 124th Wing, Boise Airport/Gowen Field, Idaho ANG

KC
303rd Fighter Squadron, 442nd Fighter Wing, AFRC, Whiteman AFB, Missouri

MA
131st Fighter Squadron, 104th Fighter Wing, Westfield Barnes Airport, Massachusetts ANG

MD
104th Fighter Squadron, 175th Wing, Martin State Airport, Baltimore, Maryland ANG

NO
706th Fighter Squadron, 926th Fighter Wing, AFRC, NAS JRB New Orleans, Louisiana

OS
25th Fighter Squadron, 51st Fighter Wing, Osan AB, Korea

OT
53rd Test and Evaluation Wing, Eglin AFB, Florida

OT
422nd Test and Evaluation Squadron, 53rd Test and Evaluation Group, Nellis AFB, Nevada

PA
103rd Fighter Squadron, 111th Fighter Wing, NAS JRB Willow Grove, Pennsylvania ANG

SP
81st Fighter Squadron, 52nd Fighter Wing, Spangdahlem AB, Germany

ST
82nd Training Wing, Sheppard AFB, TX

WA
66th Weapons Squadron, 57th Wing, Nellis AFB, Nevada

BRAC 2005 A-10 IMPACT

ACTIVE DUTY USAF

Davis Monthan AFB, AZ (66 to 66 aircraft)
No Change 66 A-10s

Eielson AFB, AK (18 to 0 aircraft)
Close 354th Fighter Wing. Move A-10s as follows:
12 Aircraft to Moody AFB, GA
3 Aircraft to 917th Wing AFRC, Barksdale AFB, LA
3 Aircraft to Backup Inventory

Moody AFB, GA (0 to 48 aircraft)
Add 48 A-10s to Moody AFB. as follows:
36 Aircraft from 23rd Fighter Group, Pope AFB, NC
12 Aircraft from 354th Wing, Eielson AFB, AK

Nellis AFB, NV (10 to 10 aircraft)
No Change 10 A-10s

Pope AFB, NC (36 to 0 aircraft)
Close 23rd Fighter Group at Pope AFB. Move 36 aircraft to Moody AFB, GA

Overseas bases (Osan AB, Korea and Spangdahlem AB, Germany are not included in BRAC)

AIR FORCE RESERVE COMMAND

Barksdale AFB, LA (15 to 24 aircraft)
917th Fighter Wing increases to 24 A-10s as follows:
15 Aircraft presently in 47th FS of 917th Wing
3 Aircraft from 354th Fighter Wing, Eielson AFB
6 Aircraft from 926th Fighter Wing AFRC, NAS JRB New Orleans, LA

NAS JRB New Orleans, LA (15 to 0 aircraft)
Close 926th Fighter Wing. Move 15 A-10s as follows:
9 Aircraft to 442d Fighter Wing AFRC, Whiteman AFB, MO
6 Aircraft to 917th Wing AFRC, Barksdale AFB, LA

Whiteman AFB, MO (15 to 24 aircraft)
442d Fighter Wing increases to 24 A-10s as follows:
15 Aircraft presently in the 442d Fighter Wing
9 Aircraft from 926th Fighter Wing AFRC, NAS JRB New Orleans, LA

AIR NATIONAL GUARD

Barnes ANGB, MA (15 to 24 aircraft)
In the USAF recommendations the 104th Fighter Wing increased to 24 keeping its 15 A-10s and receiving 9 additional A-10s from 103d Fighter Wing, Bradley IAP, CT. The BRAC Commission changed the recommendations, transferring the 18 A-10s to the 188th Fighter Wing, Arkansas ANG at Fort Smith, replacing their F-16s. The 104th Fighter Wing will convert to F-15s, receiving the aircraft from the 102 Fighter Wing MA ANG at Otis ANGB which is closing.

Bradley International Airport, CT (15 to 0 aircraft)
Close 103d Fighter Wing. Originally of the 15 A-10s assigned to the 103d Fighter Wing, 9 Aircraft were to be reassigned to the 104th Fighter Wing, Barnes ANGB, MA and 6 Aircraft were to be retired. In the BRAC Commissions recommendations, the 9 aircraft will be assigned to the 188th Fighter Wing Ft Smith, Arkansas instead.

Gowen Field, Boise Air Terminal, ID (15 to 18 aircraft)
124th Wing increases to 18 A-10s as follows:
15 Aircraft presently in the 124th Wing
3 Aircraft from 111th Fighter Wing, NAS JRB Willow Grove, PA

Martin State Airport, Baltimore, MD (15 to 18 aircraft)
175th Wing increases to 18 A-10s as follows:
15 Aircraft presently in the 175th Wing
3 Aircraft from 111th Fighter Wing, NAS JRB Willow Grove, PA

NAS JRB Willow Grove, PA (15 to 0 aircraft)
Close the 111th Fighter Wing and move 15 A-10s as follows:
3 Aircraft to the 124th Wing, Gowen Field, Boise Air Terminal, ID
3 Aircraft to the 175th Wing, Martin State Airport, Baltimore, MD
3 Aircraft to the 127th Wing, Selfridge ANGB, MI
6 Aircraft to be retired

Selfridge ANGB, MI (0 to 18 aircraft)
127th Wing will replace its F-16s with 18 A-10s as follows:
15 Aircraft from the 110th Fighter Wing, W.K.Kellogg Airport, Battle Creek, MI
3 Aircraft from 111th Fighter Wing, NAS JRB Willow Grove, PA

W.K.Kellogg Airport, Battle Creek, MI (15 to 0 aircraft)
Close the 110th Fighter Wing and move 15 A-10s to the 127th Wing, Selfridge ANGB, MI

GLOSSARY

AB	Air Base
ACC	Air Combat Command
AFB	Air Force Base
AFRC	Air Force Reserve Command
AFRES	Air Force Reserve
ANG	Air National Guard
ANGB	Air National Guard Base
ARB	Air Reserve Base
BRAC	Base Realignment and Closure
FS	Fighter Sq
FW	Fighter Wing
FWS	Fighter Weapons Sq
FWW	Fighter Weapons Wing
GP	Group
JARB	Joint Air Reserve Base
LASTE	Low-Altitude Safety and Targeting Enhancement
OG	Operations Group
PACAF	Pacific Air Forces
TAC	Tactical Air Command
TASS	Tactical Air Support Sq
TES	Test and Evaluation Sq
TEG	Test and Evaluation Group
TFW	Tactical Fighter Wing
USAFE	United States Air Force Europe
WG	Wing
WPS	Weapons Sq